Thomas Edison
Genius of Electricity

Thomas Alva Edison (1847–1931).

Pioneers of Science and Discovery

Thomas Edison
Genius of Electricity

Keith Ellis

PRIORY PRESS LIMITED

Other Books in this Series

Carl Benz and the Motor Car   Doug Nye
George Eastman and the Early Photographers   Brian Coe
Richard Arkwright and Cotton Spinning   Richard L. Hills
James Simpson and Chloroform   R. S. Atkinson
Edward Jenner and Vaccination   A. J. Harding Rains
Louis Pasteur and Microbiology   H. I. Winner
Alfred Nobel, Pioneer of High Explosives   Trevor I. Williams
Michael Faraday and Electricity   Brian Bowers
Ernest Rutherford and the Atom   P. B. Moon
Rudolf Diesel and the Diesel Engine   John F. Moon
Thomas Telford, Father of Civil Engineering   Keith Ellis
Isaac Newton and Gravity   P. M. Rattansi
Thomas Edison, Genius of Electricity   Keith Ellis
William Harvey and Blood Circulation   Eric Neil
Alexander Fleming and Penicillin   W. Howard Hughes
Marie Stopes and Birth Control   H. V. Stopes-Roe and Ian Scott
James Cook, Scientist and Explorer   Trevor I. Williams
Joseph Lister and Antisepsis   William Merrington
Alexander Graham Bell and the Telephone   Andrew McElroy

SBN 85078 126 4
Copyright © by Keith Ellis
First published in 1974 by
Priory Press Ltd., 101 Grays Inn Road, London, WC1
Text set in 12/14 pt. Photon Baskerville, printed by photolithography,
and bound in Great Britain at The Pitman Press, Bath

Contents

Illustrations

1 *The Young Inventor*

Top left Thomas's mother and father, Nancy Elliott Edison and Samuel Ogden Edison Junior.

Bottom left The house in Milan, Ohio, U.S.A. where Thomas Alva Edison was born on the 11th February, 1847. It was built by his father in 1841.

Some men have the luck to live at exactly the right time. Thomas Alva Edison was one of them. In 1847, America was largely a farming country with a population of 20 millions. By Edison's death in 1931, it had grown into the richest and most technologically advanced nation in the world with a population of 120 millions. Edison's genius for invention helped to make that growth possible.

He was born on 11th February, 1847 at Milan, Ohio. His father, Samuel Edison, was a tall, lean, jack-of-all-trades of Dutch stock. As a tavern keeper at Vienna in Ontario, Canada, he had married Nancy Elliott, a village schoolteacher. In 1837, he joined a rebellion against the Canadian government and, when it failed, fled across the border to America, settling at Milan, then a rapidly growing port on the Huron Canal. There he was joined by his wife and four children. He made a fair living as a lumber merchant.

At Milan, three more children were born. The youngest was Thomas Alva. He was an odd-looking boy with a large head, a round face, fair hair and blue eyes. He was always asking questions and he would, whenever possible, test out the answers for himself. When told that geese hatched out their eggs by sitting on them, he made a nest in the barn, put some eggs in it and sat on them for hours.

His father, Sam, thought him stupid. Both he and Mrs. Edison frequently whipped him with a birch switch, but he still got into scrapes. Experimenting with fire one day, he burned down his father's barn. Sam Edison called out all his neighbours and their children and thrashed his six-year-old son in the village square.

Milan was bypassed by a new railway which took

9

traffic from the canal and hit local trade. As a result, Sam Edison moved to Port Huron, Michigan, where he set up a lumber and grain business. Soon afterwards, he sent for his family.

Young Edison did not start school until he was eight, perhaps because of scarlet fever and other illnesses. Three months later, he walked out when his teacher told him that his brain was "addled." His mother was so indignant that she decided to teach him at home.

He made rapid progress. By ten, he was reading, or was having read to him, Gibbon's *Decline and Fall of the Roman Empire*, the *Dictionary of Sciences* and much of Shakespeare and Dickens. He seems to have had little affection for his father. "My mother was the making of me," he said later.

Without realizing it, he was also educating himself. He was passionately interested in electricity and had a simple laboratory in the cellar. Here he carried out one of his earliest experiments. Hoping to generate electricity, he tied two cats together, attached wires to their legs and rubbed their backs. The attempt failed but it is a good example of the enterprise, originality and insistence on practical experiment that was to make him one of the greatest inventors in history.

Sam Edison's business did not flourish. To help the family budget, Mrs. Edison allowed her son, Thomas, to take a job on the Grand Trunk Railroad when he was only twelve. He became a "candy butcher" on the daily train linking Port Huron with Detroit, the state capital, which then had a population of 25,000. He was not paid. His income came from profit on the sweets, popcorn and newspapers he sold to the passengers.

Each morning, he joined the train at seven a.m. and arrived at Detroit some three hours later. After replenishing his stocks, he studied books on chemistry, mechanics and manufacturing at the public library. He spent hours on Newton's *Principles*

Above Thomas at the age of fourteen, when he became a "candy butcher" on the local Grand Trunk Railroad.

Above The charge of General Ulysses S. Grant (1882–1885) at the Battle of Shiloh, Tennessee in April, 1862. It was advance news of the outcome of this battle that earned Thomas $150 profit.

but finally abandoned it. "It gave me a distaste for mathematics from which I have never recovered," he said later. At four-thirty p.m., he caught the train back to Port Huron.

He was a highly successful candy butcher. He was cheeky and sometimes gave short measure but his open intelligent face and winning manner made him popular with both passengers and railwaymen.

When two more trains were added to the run, he hired assistants to sell candy on them. He bought fruit and butter from farmers, carried them on the train without paying freight charges and sold them in the state capital. He set up a stall in Port Huron where yet another assistant sold fresh vegetables he had shipped from Detroit.

He had a sharp eye for profit. The American Civil War was now raging. By hanging around the office of the Detroit *Free Press*, he got advance news of the paper's contents. On 7th April, 1862, the Battle of Shiloh was fought with a reported 60,000 casualties.

A friendly telegraph operator wired a brief announcement to stations along the line. Placards were put up saying that further details could be obtained from newspapers available on the train. By special arrangement with the editor, Edison took a thousand copies on credit. At the first stop, he sold forty papers instead of the usual two. At the next stop, he doubled the price to ten cents and sold one hundred and fifty instead of the usual dozen. At Port Huron, he sold the rest to a stampede of customers at twenty-five cents each. His profit for the day was a hundred and fifty dollars.

A page from the Grand Trunk *Weekly Herald* which was written, printed and published by Thomas from a baggage car on the Grand Trunk Railroad.

His next venture was the *Grand Trunk Herald*, a single sheet weekly newspaper. He wrote, edited and printed it in the baggage car of the train on a second-hand press that he had bought with his Shiloh profits. It was a jumble of news, market prices, schedule changes and jokes (e.g. " 'Let me collect myself,' as the man said when he was blown up by a powder mill.") Spelling was poor. "Opisition" and "attension" were typical. But he worked the circulation up to four hundred copies at eight cents.

Bustling and inquisitive, young Thomas Edison had a finger in every pie. One day, he even drove a train. But his main interest was the electric telegraph. It was the latest means of communication and had the kind of glamour we now associate with space technology. As his Shiloh coup had proved, it could bring immediate practical benefits.

For Thomas, it had a special appeal. Perhaps because of his childhood scarlet fever, he had become deaf. From the age of twelve, he never heard a bird sing. But he could hear the telegraph. Morse code was transmitted by sending an electric current along a wire. At the receiving end, the current activated an electromagnet which drew a lever towards it. When the current was broken, the lever sprang back to its original position. At each end of its swing, it struck a screw with a sharp click. The time between each pair of clicks depended on how long the sender held down his key, or switch. A long space represented a dash, a short space a dot. Instead of the "dah-dit-dah" familiar with Morse sounders, the sound of the railway telegraph was more like "umpty-iddy-umpty."

Thomas could hear these clicks clearly. In fact, he could hear them better than people with normal hearing because he was not distracted by background noises. He made his own equipment and was soon tapping out messages to a friend along a wire strung between their homes. Eventually, six houses were linked. The system finally broke down when a wandering cow knocked down a pole, became entangled in the wire and in her panic uprooted all the other poles as well.

In 1862, when he was still fifteen, Thomas was involved in an incident which changed his whole life. Every day, the train stopped at Mount Clemens station for half an hour while extra wagons were attached. These were shunted into position. On the day in question, Thomas saw Jimmy, the 2½-year-old son of J. U. Mackenzie, the stationmaster, playing on the track. A loose wagon was rolling towards him. Edison hurled himself across its path, knocking the boy to safety. As a reward, Mr. Mackenzie offered to teach him telegraphy.

Thomas seized his chance eagerly. On four nights a week, he stayed with the Mackenzies while an assistant took over the rest of his candy route between Mount Clemens and Port Huron. This left his evenings free for instruction. It is typical of him that he arrived for his first lesson with his own set of instruments which he had made in the workshop of a friendly gunsmith.

Telegraphy at Mount Clemens consisted mainly of sending and receiving messages about the arrival and departure of trains. At the end of five months, Thomas had learned all that Mr. Mackenzie could teach him. It was enough to qualify him for a thirty-dollar-a-month job as an operator at a Port Huron bookshop which ran a public telegraph service as a sideline.

In these early days of telegraphy, operators drifted from job to job. Thomas was no exception. In the next four years, he tapped his key in almost a dozen offices from Memphis, Tennessee to Boston, Massachusetts. He lived in cheap rooms, spending his salary on equipment for his never-ending experiments. He became a brilliant operator but was repeatedly sacked for playing pranks or neglecting his work so that he could study or work on some technical

The first telegraph company in the United States—with lines from Washington D.C. to New York was set up in 1845. Within a few years, telegraph wires stretched right across the United States from the Atlantic to the Pacific. For ease of construction and repair, they would always, when crossing open country, be set up to run alongside a railway line.

problem. He was shabby and untidy in appearance. A colleague of those days said that he looked like "a veritable hay-seed," or country bumpkin.

From Port Huron, he crossed over to Canada and took a job as a night operator at Stratford Junction railway station some forty miles away. To make sure he stayed awake, he had to send a brief signal to headquarters at regular intervals. Traffic was light and he much preferred to take cat-naps so that he would be fresh for his private experiments next day. He rigged up a clock mechanism to trigger a device that sent off the signals automatically. Headquarters failed to get a reply when they called his station immediately after one of these signals. A supervisor was sent to investigate and the game was up.

Thomas was given another chance. Shortly afterwards, he was told to stop a train at his station. He should have stopped it and then telegraphed confirmation. Instead, he sent the confirmation first. When he tried to stop the train, he was too late. It had passed through.

Meanwhile, a train travelling in the opposite direction had been allowed to leave the next station down the single-track line. If the drivers had not seen each other's lights and pulled up in time, he would have been responsible for a serious accident. Realizing that his negligence was a serious offence under Canadian law, he crossed back into America.

He was still only seventeen. He worked for short periods as a telegraphist at Adrian, Michigan, Fort Wayne, Indiana, and Indianapolis before graduating to the press desk of the Western Union office at Cincinnati—all in a single year. Western Union was the giant company that ran most of America's telegraph services

By now, Thomas was a firstclass operator and could work at forty-five words a minute. He simplified his style of handwriting so that he could take down messages quickly and legibly. He was earning $105 a month.

Nothing could dampen his zest for invention. He devised electric rat-traps and cockroach killers to get rid of the vermin that infested his rooms. In Cincinnati, he wired up the trough in which railwaymen washed. As soon as they dipped their hands in the water, they got a violent shock.

More jobs followed in Nashville and Memphis, Tennessee, and in Louisville, Kentucky. He lived frugally, wearing old clothes and cracked shoes. His salary disappeared in "loans" to his spendthrift colleagues or in yet more elaborate equipment. He had all sorts of ideas for improving the telegraph. He thought up methods of boosting electric current so that messages could be sent over longer distances. He

Right The earliest telegraphic instrument built for public use was first used in 1844 on an experimental line from Washington D.C. to Baltimore, thirty-nine miles away. Alfred Vail (pictured top left) used the instrument (shown here) at the Baltimore end. In Washington, a similar one was operated by Samuel Morse (bottom left), the inventor of the "Morse" code. Benjamin Franklin (top centre) was an early researcher into the nature of electricity, and Cyrus Field (bottom right) was the power behind the Atlantic telegraph cable which linked up North America to Europe in 1851.

built a device that recorded fast, incoming messages on paper tape that could be played back at a more leisurely speed later. He used this secretly at Indianapolis when faced with a flood of high speed press messages. All went well until he fell two hours behind and complaints poured in. The manager found his gadget and banned it.

At Louisville, he was on the point of perfecting a duplex system by which two messages could be sent at the same time on a single wire but the manager forbade him to experiment with the office equipment. Thomas felt so frustrated that he threw up his job. He

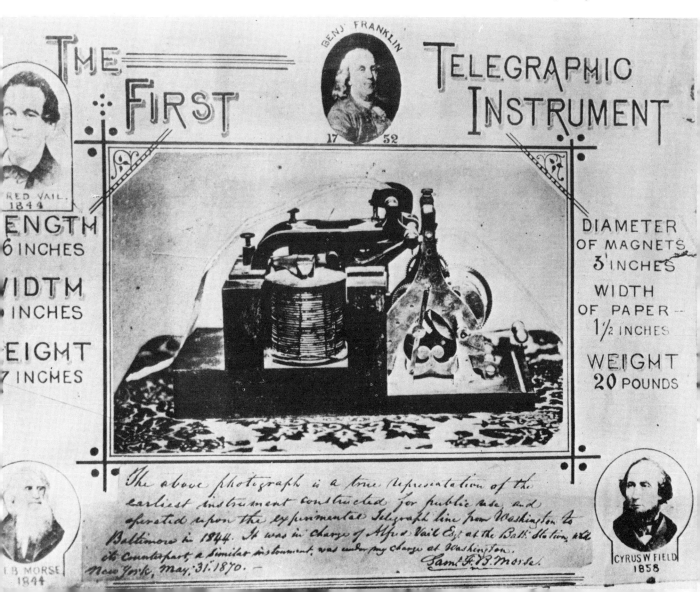

THE FIRST TELEGRAPHIC INSTRUMENT

FRED VAIL. 1844

BENJ FRANKLIN 17 52

LENGTH 16 INCHES

WIDTH 8 INCHES

HEIGHT 7 INCHES

DIAMETER OF MAGNETS 3 INCHES

WIDTH OF PAPER — 1½ INCHES

WEIGHT 20 POUNDS

The above photograph is a true representation of the earliest instrument constructed for public use, and operated upon the experimental Telegraph line from Washington to Baltimore in 1844. It was in charge of Alfred Vail Esq at the Balt Station, whilst its counterpart, a similar instrument, was under my charge at Washington.
New York, May, 31. 1870. — Saml. F.B. Morse.

S B MORSE 1844

CYRUS W FIELD 1858

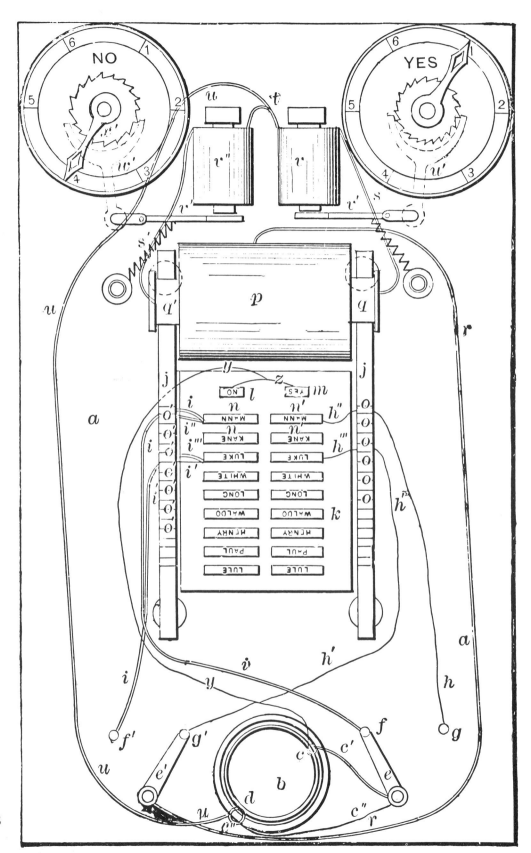

18

intended emigrating to Brazil and after a farewell visit to his parents at Port Huron, got as far as New Orleans before being put off by bad reports of the country. An operator he had known at Cincinnati then told him of a vacancy in the Western Union office at Boston, Massachusetts.

Thomas Edison was now twenty-one. He was keener than ever to work on his inventions and Boston had a light engineering industry that would increase his scope considerably. After a journey of four days and nights, he started work as a telegraphist at five-thirty p.m. on the day of his arrival.

Off duty, he had the run of a public library with more than a quarter of a million volumes. He also browsed in second-hand bookshops. "He bought," said Milton Adams, an operator with whom he shared a room, "the whole of Faraday's works on electricity, brought them home at three o'clock in the morning and read assiduously until I rose. Tom's brain was on fire with what he had read. He suddenly remarked to me, 'I've got so much to do and life is so short, that I am going to hustle.' "

The hustler had already made himself known in the many workshops that specialized in making scientific instruments. He arranged to spend his free time in that of Charles Williams, Junior, so that he could continue with his experiments. Here too he met small capitalists who were on the lookout for likely inventions to invest in. He persuaded one of these to advance him $500 in return for a half share in his duplex telegraph. Another backed his idea for an electric vote-recorder with $100. For the first time in his life, Edison had capital. He threw up his job at Western Union and became a full-time inventor.

Everything went wrong. His vote-recorder, a device by which members of Congress could vote simply by pressing a button on their desks, worked perfectly. But when he demonstrated it in Washington, the Chairman of Committees told him, "It's the last thing

Left A diagram of Edison's vote-recording machine. This machine was, in 1869, the first invention that Edison patented.

on earth we want here. Filibustering and delay in the counting of votes are often the only means we have for defeating bad legislation."

He borrowed a further $800 to develop his duplex telegraph. When it failed to work, he was left with a heavy debt.

His third main invention of these days worked and also filled a real demand. Stockbrokers needed minute-by-minute information of market prices. In New York, Dr. S. S. Laws devised a machine that showed gold prices on the dials of indicators placed in subscribers' offices. These were linked to a central transmitter by telegraph wires.

Edison's "stock ticker" was a distinct improvement. Though not the first of its kind, it showed prices not on a dial but printed out on a continuous strip of paper. It covered not just gold but any commodity or share on which information was required. It was called a "ticker" because of the noise it made.

Edison employed several men to help make the machines and soon had some thirty subscribers. Unwisely, he sold out the patent rights just as it was about to become profitable.

He decided it was time to move on again. Leaving all his equipment behind in Boston, he borrowed enough money to pay his boat fare to New York. He could scarcely have foreseen the spectacular successes that awaited him.

2 *The Robber Barons*

Edison in his early twenties.

Edison arrived in New York one morning in May, 1869. The Civil War had ended four years before and business was booming. He was literally penniless. His only assets were a small but growing reputation in the world of telegraphy and a breathtaking resourcefulness.

Wandering hungrily from the quayside, he looked through a warehouse window. Inside, a tea taster was at work. Playing the country bumpkin, Edison walked in and asked him what he was doing. "Could you spare me a packet so that I could try it for myself?" he begged. Within a few minutes, he had persuaded a nearby café proprietor to exchange his packet of tea for a cup of coffee and a plate of apple dumplings.

His hope of a bed disappeared when he found that his only friend in the city had gone away. For a day and a night, he walked the streets. Next morning he found a telegrapher friend who lent him a dollar. He lived on this for several days.

His next call was at the offices of the Gold Indicator Company, which had been formed by Dr. Laws to operate his information service for brokers. Some five hundred subscribed $300,000 a year. Frank Pope, the company's chief engineer, proved friendly. He gave Edison permission to sleep in the battery room and allowed him to spend as much time as he wished studying the complicated machinery of the indicator.

One day, it broke down. Pope was unable to locate the fault. Desperate for news, the brokers sent messenger boys who clamoured in the outer office. Dr. Laws was in a panic. Edison appeared, glanced quickly over the machinery and told him, "There is a contact spring which has broken and fallen between two cog-wheels. This prevents the gear from moving."

The fault was soon put right. Dr. Laws was so grateful that he gave Edison a job as Mr. Pope's assistant. Shortly afterwards, Mr. Pope left to form his own company and Edison became engineer at a salary of $300 a month. He soon found ways of improving the gold indicator, mainly by adding to it a device for printing out stock prices. Briefly, it worked like this:

An electric current from the central office activated two electromagnets on each of the receiving machines. One electromagnet operated a wheel embossed with type, turning it to the letter or numeral required. The second electromagnet brought the paper tape sharply into contact with this wheel. It then carried the paper forward one space ready for the next letter. The ink came from a felt roller which brushed against the rotating wheel.

The new equipment was so successful that Western Union bought out the Gold Indicator Company. General Marshall Lefferts, the new head, wanted to keep Edison on. Edison preferred to leave and set up a new company with Pope and J. W. Ashley (editor of *The Telegrapher*) specializing in electrical engineering and inventions. Pope looked after the business side and Ashley gave it free publicity in his magazine.

Lodging with the Popes at Elizabeth, New Jersey, Edison rented a tiny workshop in Jersey City. He worked harder than ever, rising at six a.m. to catch the seven a.m. train. He stayed at his bench until after midnight. The Popes' home was a half mile walk from the station. It was a bitter winter. "Many were the occasions when I nearly froze," he said.

The result of this work was a device for transmitting the prices of gold and sterling by telegraph. Subscribers could rent a machine for only $25 a week, a point which General Lefferts was quick to notice. He bought up the invention for $15,000, of which Edison received $5,000.

Left The laboratory and factory in Newark rented by Edison in 1871. From here he worked with a staff of 150 men on a large number of technical inventions, mainly to do with improvements of the telegraph.

His parents were still in Port Huron. He wrote to them:

"I. C. Edison writes me that mother is not very well and that you have to work very hard. I guess you had better take it easy after this. Don't do any hard work and get mother anything she desires. You can draw on me for money. Write me and say how much money you will need till June and I will send the amount on the first of that month. Give love to all the folks. Is Truey still with you?

Your affec son
Thos A."

His own health was robust and he regarded his deafness as an asset. He did not have to listen to small talk. "Freedom from such talk gave me an opportunity to think out my problems," he wrote in old age. "I have no doubt that my nerves are stronger and better to-day than they would have been if I had heard all the foolish conversation and other meaningless sounds that normal people hear. The things that I have needed to hear I have heard."

Edison had now been in New York little more than a year. During that time, he had patented seven separate inventions, most of them small technical improvements on the ticker. He now decided to set up on his own. Lefferts bought some of his minor inventions for Western Union, then asked him to overcome a drawback in an earlier ticker still widely used.

From time to time, it "ran wild" and printed non-sensical figures. Mechanics then had to visit every subscriber to re-adjust his machine. It might be several hours before the system was working again. Could Edison devise a method of re-adjusting all the machines at once from the central transmitter?

It took him three weeks to find a solution. Lefferts then enquired how much he wanted for his work to date. Edison had thought of asking for $5,000 and settling for $4,000. Wisely, he replied, "I'd rather you made me an offer.'' Lefferts suggested $40,000, which he accepted on the spot. At the same time, he was given an order for 1,200 stock tickers worth half a million dollars.

Edison now set up as a manufacturer. The firm was called Edison and Unger because Lefferts had insisted that he take on William Unger, a Western Union man, as partner. Edison rented workshops in Newark, New Jersey, bought machinery, hired men. His chief assistant was John Ott, who at twenty-one was three years younger than himself. Under him were two Germans, a Swiss and an Englishman, all highly skilled mechanics. The total work force was about fifty.

When meeting his business associates, Edison now wore a silk hat and frock-coat. But in the factory he dressed like a tramp. He supervised his men personally, expecting them to put in as many hours as he did himself. When a $30,000 batch of equipment developed a fault, he locked them up for sixty hours until they had tracked it down.

They put up with it because it was fun to work for him. He gave generous bonuses when things went well. Or he would suddenly take his men off on a fishing expedition. He himself spent long hours in his laboratory-cum-study in the top storey of the building. He was now taking out patents by the dozen—thirty-eight in 1872, twenty-five in 1873.

Great changes were taking place in his private life. His mother had died soon after he opened his factory and his father, now a grumbling failure, married a dairymaid.

Edison himself was attracted to one of his workers, sixteen-year-old Mary Stilwell. She was pretty and demure. Edison proposed, bought a house and married her on Christmas Day, 1871. It is said that he rushed back to his laboratory only an hour after the service and became so absorbed that he stayed until midnight. They nicknamed their first two children—Marion and Thomas—Dot and Dash.

Right On Christmas Day, 1871, Thomas Edison married sixteen-year-old Mary Stilwell. Edison is supposed to have spent the whole afternoon and most of his wedding night working in his laboratory. The illustration showing this event is rather fanciful and quite inaccurate. It shows Edison on his wedding day experimenting on light bulbs and even shows one working in the hall. He did not, in fact, start his experiments on electric lighting for at least another five years.

Edison was little interested in money for its own sake. He wanted it only so that he could buy the things he needed for his work. He had two hooks in his office, one for unpaid bills, the other for outstanding accounts. "This saved the humbuggery of bookkeeping, which I never understood," he said. "The arrangement possessed besides the advantages of being cheaper. Notwithstanding this extraordinary method of doing business, my credit was excellent."

This attitude was adequate for a twelve-year-old candy butcher, but not for a man who had to do

25

business with clever and unscrupulous financiers. He was caught between the opposing forces of Western Union and Jay Gould, a cunning, corrupt and ruthless financier who now aimed to dominate American telegraphy.

Edison was still working for Western Union when he was approached by two of Gould's henchmen—George Harrington, a former Assistant Secretary of the U.S. Treasury, and Josiah Reiff, who was what we should now call his public relations man. They were directors of the Automatic Telegraph Company, which they were secretly running for Gould. It had been formed to exploit an automatic telegraph invented by George D. Little. The invention had run into trouble and they asked Edison to put it right.

He accepted $40,000 in return for the patent rights on any inventions in connection with the automatic telegraph. He opened another factory, took Joseph T. Murray of Newark as a partner and hired Edward H. Johnson, a young railway engineer, as an assistant.

Early telegraphs had the disadvantage that only one message could be sent in one direction in a wire at any one time. The first idea for improving this state of affairs was the "duplex" which sends signals in both directions at once. Although experiments were begun as far back as 1852, the first workable version (above) was finally produced in 1872 by J. B. Stearns.

The automatic telegraph was an instrument which punched paper tape with Morse characters. This was then run through a machine which transmitted at *several thousand words a minute*. But a special type of paper was needed to record the characters at the receiving end.

One of Edison's assistants said later, "I came in one night and there sat Edison with a pile of chemistries and chemical books that were five feet high when they stood on the floor. He had ordered them from New York, London and Paris. He studied them day and night. He ate at his desk and slept in his chair. In six weeks he had gone through the books, written a volume of abstracts, made two thousand experiments and had produced a solution which would record 200 words a minute on a wire 250 miles long. He ultimately succeeded in recording 3,100 words a minute."

Edison now closed down Edison and Unger, the firm that made stock tickers. While continuing to work for the Gould interests, he also patented two copying devices on his own account, the electric pen and the mimeograph. More important, he persuaded Western Union to back his ideas for a duplex telegraph which would carry two messages at once along a single wire.

In the spring of 1873, Edison sailed to England to demonstrate his automatic telegraph. For various reasons, the Post Office did not buy the British patent rights. When he returned home in June, America was passing through a business depression.

He had always borrowed heavily to finance his work. Now, all his creditors were demanding payment. Unless he could raise money quickly, his workshops would be sold up.

The duplex was his main hope. He was already thinking of developing it into a quadruplex by which four messages could be passed at the same time. Basically, the idea was to use two currents of different strengths in each direction. There were two sending and two receiving instruments at each end. Each was adapted to respond only to a current of a particular strength flowing in a particular direction. Preliminary tests gave promising results.

Unfortunately, Western Union itself was economizing because of the depression. All through 1873, they refused to back Edison further. The following year, it became clear that they faced severe competition from the Automatic Telegraph Company using Edison's automatic telegraph. They decided to go ahead with the quadruplex, but only on the understanding that Edison should take as his partner George Prescott, Western Union's chief engineer. Prescott did none of the work but he was to receive half the proceeds. Using Western Union workshops, equipment and telegraphers, Edison soon perfected his quadruplex. He and Prescott formally agreed to sell it to Western Union.

That should have been the end of his difficulties. But they became even worse. Western Union refused to pay him for his work and creditors were threatening to sell up his home. He needed $10,000 immediately.

At this point, Western Union applied for an injunction against the Automatic Telegraph Company to stop them using the Page relay on which Western Union held patent rights. The relay strengthened signals that became weak when sent over long distances. Without it, A.T.C. would be out of business. Harrington asked Edison to devise an alternative. He did so and got $10,000 as an advance payment on the patent rights. This saved his home.

He was still desperately short of money. Western Union advanced him $5,000 which paid off his most pressing debts. Though the quadruplex would clearly save many millions of dollars by enabling one line to do the work of four, they refused to discuss further payment.

General Thomas T. Eckert, the general superintendent of the Western Union Telegraph Company. He shortly became president of the rival Atlantic and Pacific Telegraph Company after arranging a deal in which this company bought from Edison his rights in the quadruplex system.

General T. T. Eckert was then general superintendent of Western Union. He heard of Edison's plight and told Jay Gould who immediately offered Edison $30,000 as an advance for his share of the quadruplex rights. A tenth of the profits made by the quadruplex would follow. Exasperated by Western Union's delays, Edison accepted. Eckert defected to Gould who now came into the open as the power behind A.T.C. He bought it up and merged it with his Atlantic and Pacific Telegraph Company, promising to make Edison chief electrician. Edison's share of the quadruplex profits was to come from Atlantic and Pacific stock. The job did not materialize. The stock proved worthless. After a final protest to Gould, Edison told him that he would work henceforth only for Western Union which "was only too glad to get him back."

Western Union asked the courts to cancel Edison's agreement with Gould so that they alone could use the invention. Their lawyers denounced Edison as a "professor of duplicity and quadruplicity" who had bitten the hand that fed him. Yet only four years later, the president of Western Union admitted in his annual report that the invention was "one of the most important in telegraphy and had saved the company $600,000 annually."

Edison spent his $30,000 advance from Gould in developing an octuplex, which he hoped would carry eight messages simultaneously. "This," he said, "I never completed." He now gave up manufacturing and left the world of finance to the robber barons. Wisely, he returned to his true vocation, industrial invention.

From now on, he would head teams of experts and so be able to work on many projects at the same time. Invention would no longer be the monopoly of lonely geniuses. It would become the concerted effort of scores, perhaps hundreds, of well trained scientists and technicians. In an advanced, technological society, this must always be the main pattern. And it was Edison who first set it.

Edison's "phonoplex" (multiplex) receiving unit.

3 Disembodied Voices

The commissioner of the New York Patent Office once referred to Edison as "that young man in New Jersey who has made the path to the Patent Office hot with his footsteps." He might be working on as many as forty-five devices at the same time. He sold them for large sums of money which he spent almost immediately on further experiments.

Harassed by the visitors that thronged his New Jersey workshops, he built new premises at Menlo Park, a small hamlet in remote, lightly-wooded country twenty-four miles from New York City. "When the public tracks me out here," he said, "I shall simply have to take to the woods."

The Menlo Park laboratory in the winter of 1879.

His father came down from Port Huron to superintend the building. The main workshop was 100 feet long and 35 feet wide. It had two storeys. The ground floor housed lathes, punches, drills and milling machines powered by an eighty horsepower engine, as well as a wide variety of electrical apparatus. The first floor was an experimental laboratory with more electrical equipment, microscopes, air pumps and every type of chemical and mineral that might be needed. There was a large library of the latest science books.

"The wizard of Menlo Park" was a new type of inventor. He had little in common with either the cranky individualist of legend or the academic interested only in science for its own sake. He was wholeheartedly commercial. He might be the inspirer but he was also the head of a team. Menlo Park was probably the first industrial research laboratory in the world.

The nucleus of his team was the band of technicians he had first enrolled at Newark. Edward Johnson looked after the business side. In 1878, Edison hired Francis R. Upton, a Princeton graduate who had studied under the eminent German scientist Hermann von Helmholtz, to work on mathematical problems.

They put in long hours. They were keen to do so because they enjoyed working for Edison. He had a nickname for everyone. Upton's, for intance, was "Culture." Together with the Edison family and their three negro servants, the laboratory staff soon outnumbered the other inhabitants of Menlo Park. It became known as "Edison Village."

It is impossible to describe all the inventions that poured out. Many were technical improvements on the telegraph. A tasimeter measured heat so precisely that it reacted to a gas jet a hundred feet away. An odorscope measured smells that were otherwise undetectable. A tripod-mounted megaphone fitted with

Below Thomas Edison (centre) with his chief Menlo Park assistants.

Above Elisha Gray's telephone. Elisha Gray and Alexander Graham Bell both went to the Patent Office on the same day to describe almost exactly the same invention. Bell won the race by a few short hours.

twin trumpets and ear tubes enabled people to speak to each other across several miles.

Two groups of inventions dominated Edison's attention during these years—those concerned with the telephone and phonograph, and those to do with electric light.

Edison did not invent the telephone. That honour clearly belongs to Alexander Graham Bell of Salem, Massachusetts. He filed the patent on 14th February, 1876, beating a rival inventor, Elisha Gray of Chicago, Illinois, by only a few hours.

Bell's telephone had an iron disc or diaphragm which vibrated when someone spoke into it. Behind the disc was a magnet surrounded by a coil of wire. Movements of the disc set up currents of electricity in the coil of wire. This is called electromagnetic induction.

The electricity flowed along a wire to the receiver at the other end. It passed into a coil of wire and varied the strength of a magnet inside. The varying strength of the magnet set up vibrations in an iron disc. These vibrations corresponded to those of the first iron disc and so were heard as speech.

Bell's telephone had serious drawbacks. There was only one instrument at each end and this had to be used for both sending and receiving. Even when shouted, messages could be heard only faintly and were often drowned by hisses and crackles. Maximum range was about two miles. Even so, there was widespread public interest. Frightened of competition, Western Union asked Edison to devise an alternative.

The job was especially difficult for him because of his deafness. He often had to ask colleagues to listen for him, though sometimes he could "hear" by feeling the vibrations with his teeth. On the other hand, deafness had its advantages. It made him work to get a loud, clear sound which even he could hear.

He had a reputation for solving every problem put to him. The telephone was no exception. He used a battery to boost the strength of the signal and linked it with a transformer which stepped it up even further. The necessary *variation* in strength had to come from a variable resistance which somehow had to correspond to the sound waves of the human voice.

Edison sometimes thought out the solution to a problem. At other times, he worked by trial and error. Now, he was looking for a substance whose ability to carry electricity would vary according to the pressure placed upon it by the vibrations of the iron disc. He tried some 2,000 different chemicals before finding the most suitable—carbon. He placed it in the form of a button between two metal plates behind the metal disc and incorporated it into the circuit. He patented his transmitter, really a microphone, in February, 1878.

Right Edison's loud-speaking telephone, developed in 1878 and 1879.

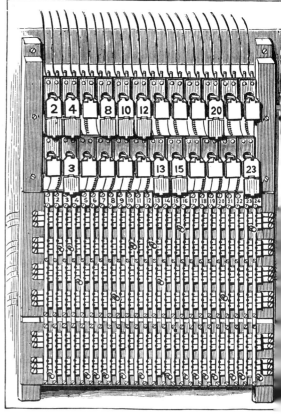

1. Switch Board and Telephone. *a*. Receiver. *b*. Trans to rotate the Chalk Cylinder.
2. Part of Switch Board (larger scale).

Left An audience in Boston Library listening to a telephone speech by Alexander Graham Bell. *Right* A full-scale model of Bell's telephone, patented in 1876.

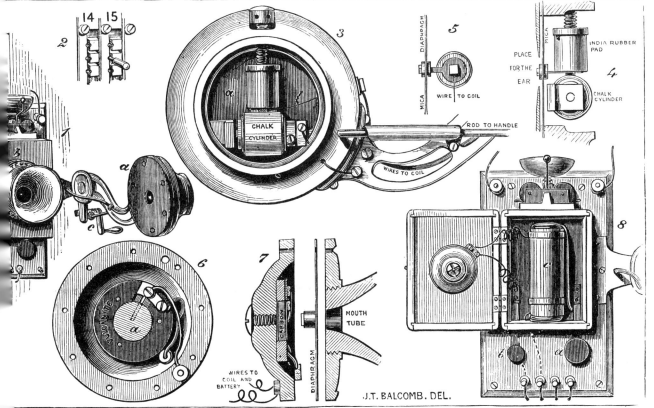

J.T. BALCOMB. DEL.

andle
3. Inside of Receiver. *a*. Diaphragm.
4. Another view of Chalk Cylinder.
5. Metal Slip and Platinum Knob.

6. Part of Carbon Transmitter. *a*. Carbon between two metal discs.
7. Section of Transmitter, with additions.
8. Transmitter Apparatus opened.

It was an immediate success. It transmitted the sound of a voice loud and clear over hundreds of miles. But it *was* only a transmitter. It had to be used in conjunction with a Bell receiver. Even so, Western Union bought the rights for $100,000.

Two rival networks were set up—the Bell System based on Boston and the American Speaking Telephone Company, a subsidiary of Western Union, based on New York. The legal position was complicated. In essence, Bell owned the receiver and pirated Edison's transmitter. The American Speaking Telephone Company owned Edison's transmitter and pirated Bell's receiver.

It was an impossible situation. Western Union decided to leave the telephone field to Bell. They handed over Edison's transmitter and their own telephone network in return for payments which eventually came to $3.5 million. Edison also did well. Foreign patent rights and further development work brought him some $250,000.

Even before completing his work on the transmitter, he had invented an almost equally important device that no-one else had ever even thought of, the phonograph, a forerunner of the gramophone.

The idea came to him while working on a repeater for messages in Morse code. This was a continuous tape of paraffined paper wound round a grooved cylinder. A chisel-shaped stylus made indentations on the paper corresponding to dots and dashes. As in an ordinary telegraph, the movements of the stylus were controlled by the making and breaking of an electric circuit by the operator. The tape could then be run through another machine with a stylus that bounced up and down on the indentations. In doing so, it alternately made and broke a circuit, thus sending out Morse signals.

"I found that when the cylinder carrying the indented paper was turned with great swiftness," wrote

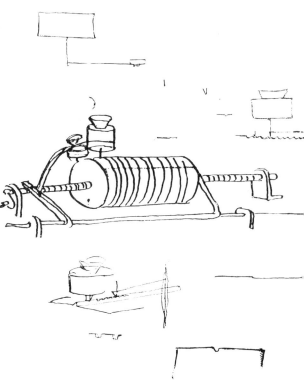

Edison's original sketch of the phonograph.

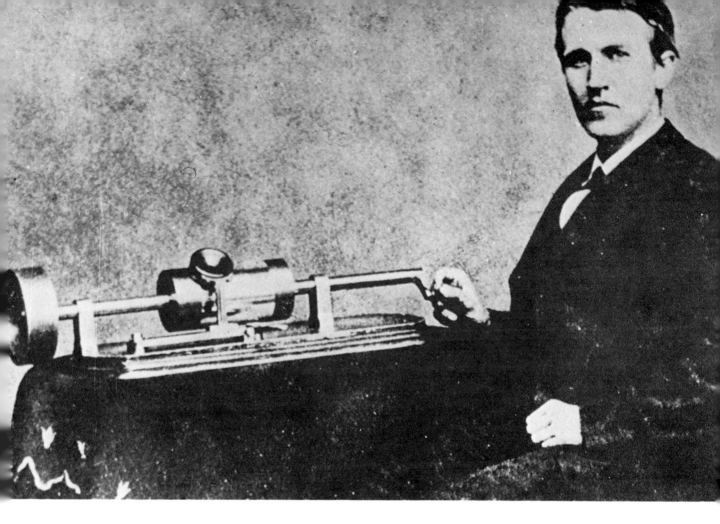

Edison with his phonograph in 1878 on its first public showing.

Edison, "it gave off a humming noise from the indentations—a musical, rhythmic sound resembling that of human talk heard indistinctly."

Edison had an eye for happy accidents. He wondered if human talk really could be reproduced.

The iron disc of the telephone transmitter turned sound waves into vibrations. He fitted it with a stylus touching a strip of paraffined paper running underneath. "I shouted the words 'Halloo! Halloo!' into the mouth-piece, ran the paper back over the steel point and heard a faint 'Halloo! Halloo!' in return. . . . That's the whole story."

John Kruesi, his best mechanic, was given the job of making a working model at a piece-work price of $18.

It consisted of a shaft carrying a grooved cylinder covered with tin foil. As the shaft was turned by a hand crank, the cylinder travelled across the machine. On one side of it was a recording diaphragm and stylus, on the other a reproducing stylus and diaphragm. Each could be swung into contact with the cylinder as needed.

When it was ready, the staff gathered round for a demonstration. The foreman bet Edison a box of cigars that it would not work. Edison slowly turned the handle and repeated the nursery rhyme "Mary had a little lamb" into the recording diaphragm. Then he wound the cylinder back, swung the reproducing stylus into position and turned the handle again. His voice was heard clearly repeating the nursery rhyme. "Well," said the foreman, "I guess I've lost."

Edison successfully applied for a patent on his phonograph on 15th December, 1877. It would, he said, "be largely devoted to music, either vocal or instrumental—and may possibly take the place of the teacher. It will sing the child to sleep, tell us what

Above The phonograph in London. Professor John Tyndall of the Royal Institution demonstrated the phonograph with the help of an assistant to an audience which included Alfred Lord Tennyson. Professor Tyndall recorded a short piece from Tennyson's poem, "Come into the garden, Maud."

The phonograph recording a cornet *right* and a grand piano *below*.

o'clock it is, summon us to dinner, and warn the lover when it is time to vacate the front porch. As a family record, it will be precious, for it will preserve the sayings of those dear to us and even receive the last messages of the dying. It will enable the children to have dolls that really speak, laugh, cry and sing, and imitation dogs that bark, cats that meow, lions that roar, and roosters that crow. It will preserve the voices of our great men and enable future generations to listen to speeches by a Lincoln or a Gladstone. Lastly, the phonograph will perfect the telephone and revolutionize present systems of telegraphy."

Edison's phonograph brought him world fame. He was invited to demonstrate it to President Rutherford Hayes in the White House, the President's official residence in Washington, D.C. The crowds who now bore down on Menlo Park were treated to a repertoire of tricks. Recorded songs were interrupted by cries of "Help! Police! Murder!" By spacing out snippets of repartee on the cylinder, Edison was able to hold a conversation with the machine as he turned the handle.

A company was formed to manufacture phonographs. Edison was paid an advance of $10,000 against a royalty of 20 per cent on every machine sold. They were set up in arcades where masses paid to hear them. But not for long. Reproduction was poor. The cylinders played for little more than a minute and quickly wore out. Edison himself was disappointed that it was not good enough to use as an office dictating machine. The phonograph, it seemed, was just a passing novelty. He turned his attention to other things.

Top and bottom right In 1878 an arc light was set up in the clock tower at the Houses of Parliament to burn for as long as the House of Commons was in session. Each pair of carbon points lasted for five hours before they burnt out.

Near right One of the many uses that Edison saw for his phonograph was as an office dictating machine.

4 Let There Be Light

Though still only thirty-one, Edison was now a national figure. Strongly built and of medium height, he had a square, serious face and a neat quiff which reminded people of Napoleon. He usually wore a business suit which, except in studio portraits, was rumpled and even shabby. This is presumably because of the long hours he worked, rarely sleeping more than six hours a night and often making do with cat-naps at his desk.

In the summer of 1878 he went on a rare vacation to the Rocky Mountains with a part of scientists. The most important outcome was a suggestion by his friend Professor George Barker that he take up the problem of electric light.

It was not a new idea. Sir Humphry Davy, the English chemist, had demonstrated an arc light as early as 1808. This was produced when a powerful current of electricity leapt across the gap between two carbon points. In 1831, Michael Faraday, another English chemist, invented a dynamo which generated electricity in sufficient quantities to keep these lights going. Since then, arc lights had been used in English lighthouses, Paris streets and even a few American stores.

"At the time," wrote Edison, "I was more or less at leisure, because I had just finished working on the carbon button telephone, and this electric-light idea took possession of me. It was easy to see what the thing needed: it wanted to be sub-divided. The light was too bright and too big. What we wished for was little lights, and a distribution of them in people's houses in a manner similar to gas."

From the very beginning, then, Edison was thinking on a vast scale. He aimed at nothing less than a complete system of electric lighting—a central power station, a generating plant, mains and branch lines to distribute electricity, safety fuses and electric bulbs in every room of every house, together with meters to measure how much current had been used. He was challenging the huge American gas industry which gathered in some $150 million a year, mostly by providing home and business lighting in the big cities.

It was easy to subdivide the power. Instead of connecting up the lights "in series" (i.e. in one continuous chain), he would wire them "in parallel" (i.e. each would have its own separate leads to the negative and positive wires from the generator). If a lamp failed or was switched off, it would not break the entire circuit, plunging whole areas into darkness. That lamp alone would go out.

Inventing a low-powered electric light was much more difficult. What was needed, said Edison, was "a candle that will give a pleasant light, not too intense, which can be turned on and off as easily as gas. Such a candle cannot be made from carbon points, which waste away, and must be regulated constantly while they last. Some composition must be discovered which will be luminous when charged with electricity and that will not wear away."

He was not thinking of getting light from an electric arc but from a substance which gave a bright glow when heated by electricity. This is called an incandescent lamp.

On 5th October, 1878, Edison applied for a patent on an incandescent lamp with a platinum filament. It fulfilled all his requirements, except that the platinum melted after ten minutes and the lamp had to be replaced. Nevertheless, on the advice of his lawyer, Grosvenor Lowrey, he announced that the problem was largely solved. Moreover, his system would make available electric power for everything from cookers to sewing machines.

His remarks were reported all over the world.

A more usual place for arc lights was in the street. This house in New York in 1879 also used them inside. Not many houses did, though, as they were generally thought to be much too bright.

Many scientists were scornful. But, as Lowrey had guessed, American financiers took more notice. Gas shares fell. The Edison Electric Light Company was set up to back him. It gave him $50,000 for research and a large block of shares. In return, it was to have sole rights in his electric lighting inventions over the next five years. The directors were mainly Western Union directors, but J. Pierpont Morgan, the banker, had a large finger in the pie.

Rashly, Edison had predicted that he could produce a system in six weeks. It took him well over a year. New buildings went up at Menlo Park—an

Opposite Edison and his assistant
experimenting with electric light in
the late 1870s.

engine house, a glassblower's shed and an office
which also housed a library and a hospitality room for
distinguished visitors.

The light bulb was still the main problem. When
heated to the point where it gave off a bright light,
platinum melted. Carbon, which had a melting point
of 3,500 degrees centigrade, burned out long before
the right temperature was reached, even in a vacuum.

He tried dozens of other materials, nearly blinding
himself with the brief flash given off by nickel. Then
he tried platinum again. The newly invented Sprengel
pump enabled him to obtain a much higher vacuum
than was previously possible. This had several effects.
Any material lasted longer before burning out. Also,
it was possible to extract from the incandescent
materials hidden gases trapped inside them. This
made them still more resistant to burning out. It

Left and right Early experimental
Edison lamps.

44

hardened them, too, so that they were less liable to melt. By April, 1879, he had a much brighter platinum light. But it still fused far too quickly. He now intensified his search for a suitable alternative. Some 1,600 materials were tested.

Meanwhile, "Culture" Upton was largely responsible for calculating the size (and therefore the cost) of the copper mains required to carry the amount of electricity needed to make hundreds of thousands of lamps light up.

At the same time, Edison was working on an improved dynamo. He needed one which would provide a constant voltage for all parts of his system. It also had to be economical. By the middle of 1879, he had found the type of dynamo he required. It was driven by steam. Instead of the usual belt drive, which wasted a good deal of energy, the shaft of the steam engine itself drove the dynamo. Its voltage was a fairly steady 110.

Edison went back to the bulb. None of the materials tried had proved successful. New calculations by Upton showed that the lamp had got to use even less power than he had previously thought. He needed a material, therefore, with a high electrical resistance. For maximum economy, it had to be only one sixty-fourth of an inch thick and six inches long.

Theoretically, carbon was ideal but, as we have seen, it tended to burn out. His new technique for driving out hidden gases made this less likely. He experimented with carbon again, this time in the form of soot, mixed with tar and moulded by hand into fine threads or filaments.

Tension rose high. His money was running out and his backers were becoming impatient. He was now sleeping less than four hours a day. When his men flagged, he ordered food and wine for a party. They sang or listened to recordings of comic songs on the phonograph.

Edison's dynamo of 1879 (called the "Long-waisted Mary Ann") was driven by steam engines in the room beyond. It was an improved version of one built by Werner von Siemens in 1866 in Germany. When run at the correct speed, Edison's dynamo was able to produce a fairly constant voltage of 110.

MACHINE SHOP.

47

The soot filaments burned up to two hours. He felt he was on the right track and looked for other forms of carbon. The breakthrough came from a reel of ordinary sewing cotton. He placed a hairpin-shaped length in a nickel mould, heated it in a furnace for five hours and carefully took out the thread of carbon that remained. It broke immediately.

Edison persevered. "All night, Bachelor, my assistant, worked beside me," he wrote. "The next day and the next night again, and at the end of that time we had produced one carbon out of an entire spool. Having made it, it was necessary to take it to the glass-blower's house. With the utmost precaution, Bachelor took up the precious carbon, and I marched after him, as if guarding a mighty treasure. To our consternation, just as we reached the glass-blower's bench the wretched carbon broke.

"We turned back to the main laboratory and set to work again. It was late in the afternoon before we had produced another carbon, which was again broken by a jeweller's screw-driver falling against it. But we turned back again, and before night the carbon was completed and inserted in the lamp. The bulb was exhausted of air, and the sight we had so long desired to see met our eyes."

It burned for forty-five hours on 11th and 12th October, 1879. On 1st November, Edison applied for a patent. Yet he was still not satisfied. The cotton filament had shown that vegetable fibres were most suitable. Was there one even more effective than cotton? He tried every type he could lay his hands on from boxwood to human hair. The best turned out to be cardboard. By Christmas, Menlo Park was ablaze with lights with burnt cardboard filaments.

After months of secrecy, Edison gave the newspapers a full story. It was a world sensation. Inventors in England, America and many other countries, who had also been working on incandescent lamps, acknowledged defeat. The business world was

Edison carbonizing a paper lamp filament. Months of experimenting eventually resulted in the choice of a Japanese bamboo as the best source for carbonized lamp filaments.

Above Edison's lamp factory at Menlo Park in 1880.

equally impressed. Shares in the Edison Electric Light Company sextupled in value. More important, the directors, who had been uneasy at the long delay, now advanced him another $57,000 for development.

Visitors thronged to Menlo Park. On New Year's Eve alone, three thousand came to wonder at the lights. One of them wrote, "The lamps are about four inches long, small and delicate, and comely enough for use in any apartment. They can be removed from a chandelier as readily as a glass stopper from a bottle and by the same motion. The current is turned on and off by pressing a button."

Not all the visitors were friendly. William E. Sawyer, a rival inventor, drunkenly accused Edison of

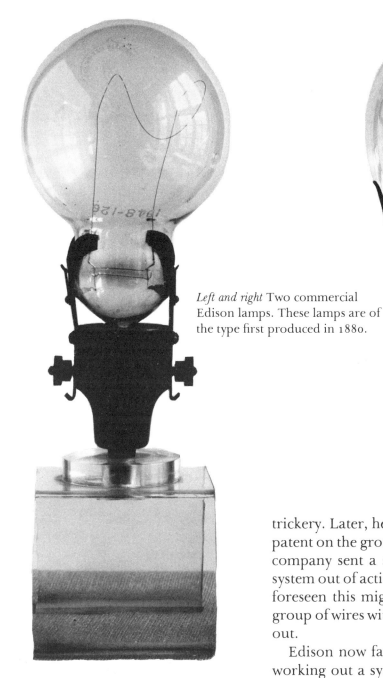

Left and right Two commercial Edison lamps. These lamps are of the type first produced in 1880.

trickery. Later, he unsuccessfully challenged Edison's patent on the ground that he had filed it first. One gas company sent a saboteur who attempted to put the system out of action by short-circuiting it. Edison had foreseen this might happen and had protected each group of wires with a safety fuse. Only four bulbs went out.

Edison now faced a host of technical problems in working out a system of distributing electricity from the dynamos in the central generating station. One was economic. "Mains" are the main cables carrying power from the central station to local distribution points. If they were big enough to carry power to 10,000 bulbs, they would cost some quarter of a million dollars in copper alone. Even then, the lights furthest from the generators would be a third less bright than those nearest because of power losses.

He solved both problems at once with his "feeder and main" system. This consisted of feeder wires which took power directly from the dynamos to distribution points on the mains. It cut the cost of copper needed by almost nine tenths. It also kept the power constant throughout the system.

This was only one of dozens of entirely new ideas that poured out of Menlo Park. Edison wanted a lamp filament that lasted longer than the 300 hours expected from burnt cardboard. He tested more than 6,000 vegetable fibres.

"Finally, I carbonized a strip of bamboo from a Japanese fan, and saw that I was on the right track. But we had a rare hunt finding the real thing. A man went down to Havana, Cuba and the day he got there he was seized with yellow fever and died in the afternoon. I sent a schoolmaster to Sumatra and another fellow up the Amazon. William H. Moore, one of my associates, went to Japan and got what we wanted there. We made a contract with an old Jap to supply us with the proper fibre, and that man went to work and cultivated and cross-fertilized bamboo until he got exactly the quality we required."

In 1880, Henry Villard, a financier who was soon to join the board of Edison Electric, gave Edison a chance of demonstrating his system to a still wider public. Villard owned the *S.S. Columbia*, a ship newly-built to link New York with the Pacific Coast via Cape Horn. (The Panama Canal had not been built.) Edison installed a complete lighting system for the ship with steam engines, dynamos and 115 lamps. It was a complete success.

Other trials were more exacting. At Menlo Park, Edison laid eight miles of underground mains, solving as he did so the problem of insulating them against loss of power. He wired the handful of houses

Edison and his assistants in the
Menlo Park laboratory some time
in the 1880s.

in the hamlet and erected lamp-posts along a network
of "streets" over half a square mile. When he flung a
switch in the central power station, the whole area
blazed with the light of 425 lamps.

By now his eyes were on New York. In January,
1881, he successfully demonstrated a huge new
dynamo powered by a 120-horsepower steam engine.
How did he know how much current would be
needed?

"Simplest thing in the world," he said. "I hired a
man to start in every day about two o'clock and walk
through the district noting the number of gas lights
burning in the various premises; then at three o'clock
he went around again and made more notes, and at
four o'clock and up to every other hour to two or
three o'clock in the morning. In that way, it was easy
enough to figure out the gas consumption of every te-
nant and of the whole district; other men took other
sections."

He believed that electricity would replace gas for lighting purposes. It was more convenient and just as cheap. The directors of Edison Electric were less sure. They refused to finance factories to manufacture the necessary equipment. So Edison manufactured it himself. For capital, he either sold his stock in Edison Electric or used it as security for loans.

Three new companies sprang up. At Menlo Park, the Edison Lamp Company was soon producing a thousand light bulbs a day. In New York, the Edison Machine Works turned out dynamos in an old dockland factory. He also had a third share in another company which made smaller items of equipment from fuses to lamp sockets. Together, these companies employed a thousand men.

By now the directors of Edison Electric were beginning to wake up. Perhaps, after all, they could beat the gas companies. Impressed by the thoroughness of Edison's market research and profit forecasts, they got permission from New York City to set up an electric lighting system. They formed a subsidiary, the Edison Illuminating Company of New York, and raised $80,000 to help finance a central station.

Edison moved into a New York hotel with his wife and three children. He rented 65, Fifth Avenue, a luxurious town house, for business purposes and quickly had it blazing with electric light. The first district he planned to supply lay to the north of Wall Street, the financial sector. He chose it because it contained every type of property from hovels to huge business offices.

"I planned out the station and where it ought to go," he said, "but we could not get real estate [land] where it was wanted. It cost us $150,000 for two old buildings down in Pearl Street where we finally settled."

No-one had ever set up a central station before. He encountered unexpected difficulties in getting the

Above Edison's New York central station for electric lighting in 1882.

Below The dynamo room of Edison's central station in New York.

54

steam engines to work together. He had to invent from scratch fuse wire, voltmeters and numerous other items which we now take for granted.

Meanwhile, there were eighteen miles of ditches to dig and mains to lay. "I saw every box poured and every connection made on the whole job," he said. "There was nobody else who could superintend it. I used to sleep nights on piles of pipes at the station."

At three p.m. on 4th September, 1882, the lights were finally switched on. There were only 400 of them. It soon became evident that the $600,000 spent so far would not reap quick profits. Bulbs cost a dollar each. Faults often brought blackouts. Some houses caught fire. In one, a man was burned to death. But Edison had shown that his lighting system worked. It was only a matter of time before electric light made gas light as obsolete as the stagecoach.

The first central lighting station was not, in fact, the New York one. Earlier in 1882 the one shown below had gone into operation in Appleton, Wisconsin, supplying power to fifty lights.

5 Man and Myth

In his lifetime, Edison became a legend. He was world famous. On several visits to England and Europe, he was fêted by the leading scientists of the day. The King and Queen of Italy gave him letters of commendation. He was entertained at the Mansion House by the Lord Mayor of London. He recorded the voices of Prince Otto von Bismarck, the German chancellor, of Lord Tennyson, the English poet, and of William Gladstone, the British prime minister.

After listening to his phonograph, Gladstone recorded this message: "I am profoundly indebted to you for, not the entertainment only, but the instruction and the marvels of one of the most remarkable evenings which it has been my privilege to enjoy. Your great country is leading the way in the important work of invention. Heartily do we wish it well; and to you, as one of its greatest celebrities, allow me to offer my hearty good wishes and earnest prayers that you may long live to witness its triumphs in all that appertains to the well-being of mankind."

But the second half of Edison's life was something of an anti-climax. Many of his large business interests never wholly fulfilled their promise. He was continually plagued with lawsuits over patent rights. He was responsible for a very few major inventions and many smaller ones. But he did not follow through his successes. He had many costly failures. He had been a pioneer in the development of the multiplex telegraph, the microphone, the phonograph and electric lighting. Now he was repeatedly overtaken by men who saw much better than he the trends which were to transform our civilization.

This was all the more surprising as he had always

Edison experimenting with "micrography"—the photography of things seen down a microscope, and greatly magnified.

been not only enterprising, but commercial. He did not try to invent useless or impossible gadgets. Indeed, he was scarcely an original inventor at all. Up to now he had kept abreast of developments himself in a dozen different fields and had built on these to invent devices that would be of use to large numbers of people. He had gone into manufacturing only because industrialists did not always share his vision. He could not bear to see a useful invention lying idle.

He did everything on a vast scale. So it is not surprising that he became a millionaire many times over. He could have retired at thirty. Instead, he preferred to keep on working until he was almost eighty.

Money in itself meant little to him. He used it to finance the work in hand. As a telegraphist, he had plunged his wages on the best experimental equipment he could afford. Now that he was rich, he invested millions of dollars in his projects. Setbacks rarely upset him. At sixty-seven, he saw seven of his factories burn down. As they were under-insured, he lost something like a million dollars. Within thirty-six hours, he had 1,500 men rebuilding still bigger factories under his personal supervision. "No-one's ever too old to make a fresh start," he said.

He never lost the common touch, nor his gift for inspiring others to work for him. He was a brilliant story-teller. He was a compulsive practical joker. He carried on chewing tobacco and spitting the juice on the floor, even when he was a multi-millionaire. When he had a problem to solve, he worked day and night with only brief snatches of sleep for days on end. His "insomnia squad," as he called his favourite workmen, never refused to work with him.

He paid good but not exceptional wages. He had no sympathy with trade unions. Eighty men whom he had trained to make light bulbs threatened to strike if he sacked an unsatisfactory workman. Edison needed them. So he kept the man on. But he secretly invented machinery for making the bulbs. When it was ready, he sacked the workman and his colleagues went out on strike. "They have been out ever since," he said grimly.

Edison was ill at ease in the world of high finance. He looked on money as a means and could never understand men who regarded it as an end. He was over-trusting. He was outmanoeuvred repeatedly by such money geniuses as the banker John Pierpont Morgan. He was swindled by individuals, often for huge sums. He once entrusted a lawyer with some applications for patents. Instead of registering them in Edison's name, the lawyer sold them to rivals and decamped with the proceeds.

Edison was a good judge of men in the sense that he repeatedly built up teams who successfully carried through his projects. He seemed little interested in people at a deeper level. He had numerous acquaintances but few real friends. One of the few exceptions was Henry Ford, the motor car manufacturer.

His flair for publicity was exceptional. He cultivated reporters and gave them sensational stories about his inventions, often before he had perfected them. He had a fund of anecdotes about himself that became steadily more colourful over the years. He delighted in sitting for the camera. He had studio portraits taken regularly and was always eager to pose for press photographers against a background of impressive equipment, peering intently into a test tube.

He "got away with it" because he always fulfilled his promises, at least in his early years. His countrymen liked him too because he was a prime example of the American dream, a poor country boy who made good. He was plain-spoken, unassuming, eager to poke fun at intellectuals and always ready, even when rich, to pitch in with his men, covering himself with grease, when necessary, and even snatching sleep on a bench.

Yet he never curried favour with his friends, his

In 1894, Thomas Edison started to make talking or "phonographic" dolls which "do not merely gurgle and squeak, but say 'Mamma' and 'Papa' in a real woman's voice, and tell fairytales and sing songs like real live boys and girls." The girl in the centre of the picture is recording the "doll's" speech onto a small cylinder which is then fitted to the miniature phonograph shown at *bottom right. Bottom left* shows the inside of Edison's phonographic doll factory.

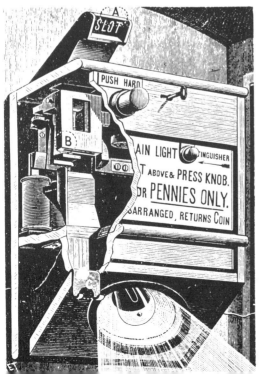

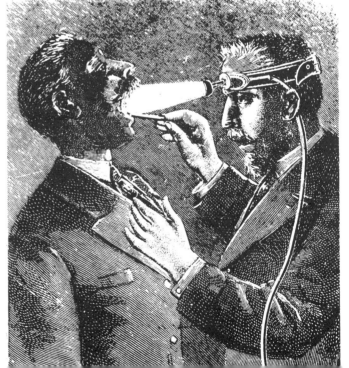

business associates or the American public. He caused a furore by announcing that he did not believe in a personal God, only a "Supreme Intelligence." But he admitted the possibility of survival after death. He repeatedly attacked mindless fact learning and preferred teaching through play because it made learning a pleasure. He was an early opponent of capital punishment. "There are wonderful possibilities in each human soul," he said. "I cannot endorse a method of punishment which destroys the last chance of usefulness."

Edison had his faults. He was sometimes insensitive to other people's feelings. His practical jokes were often far from funny. His passion for work led him to neglect his own children, especially those by his first wife. When driven into a corner, as in the argument over the relative merits of alternating current and direct current, he could hit below the belt.

Was he a genius? He once defined genius as "one per cent inspiration and ninety-nine per cent perspiration." By that standard, he clearly qualified. But most of all, he was himself. In both success and failure, Edison was his own man.

6 *Losing Touch*

In 1882 the world was Edison's oyster. He seemed all set to witness the "triumphs" of his inventions. What happened?

Edison's triumphs were slow in coming. Financiers expected a quick, sure return on their capital but central stations were costly and income from them low. After two years, Pearl Street had only 500 customers burning some 10,000 lamps. Morgan and the other major shareholders in Edison Electric preferred to invest in railways, rather than expand the New York lighting system.

There was, however, a steady demand for independent plants both in large buildings and small towns without gas-light. These were supplied by the Edison Company for Isolated Lighting and financed through the T. A. Edison Construction Department—both subsidiaries of Edison Electric. Edison's own manufacturing companies were paid for equipment partly in cash and partly in shares in the various power companies set up. They also gave a good deal of credit.

Naturally, the manufacturing companies prospered. In late 1884, Edison successfully fought a move by Edison Electric to take them over. Three years later, the directors of the parent company took the plunge. They agreed to an expansion of Pearl Street and the building of two more central stations in New York. When they came into operation two years later, there were fifty-eight central stations in other cities as well as some five hundred isolated plants.

Edison was rich, drawing dividends from both the manufacturing companies and from Edison Electric in which he still had a minority holding. But he still

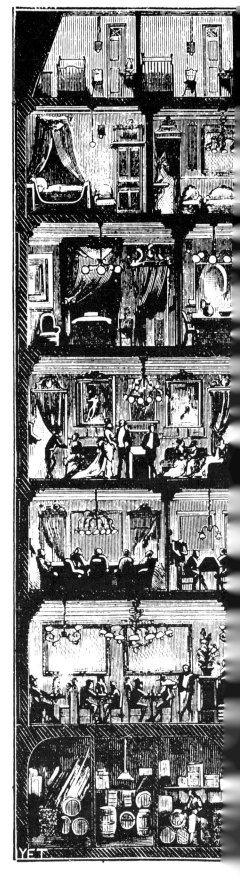

The fashion for electricity in the home caught on quickly. This house, in 1884, had an electric lift as well as electric lights in every room.

supervised the work personally. He travelled extensively and was always ready to dirty his hands and his business suit helping his workmen in an emergency.

What he hated were the endless lawsuits in which he became involved. Other inventors—some of them honest, others pirates—repeatedly tried to upset his innumerable patents by claiming that they had thought of them first. Or Edison Electric might itself be fighting rival companies infringing Edison patents.

After giving evidence in one of these cases, Edison wrote: "Waited one hour for the appearance of a lawyer who is to cross-examine me on events that occurred eleven years ago. Went on stand at 11.30. He handed me a piece of paper with some figures on it, not another mark. Asked in a childlike voice if these were my figures, what they were about and what day eleven years ago I made them. This implied compliment to the splendour of my memory was at first so pleasing to my vanity that I tried every means to trap my memory into stating just what he wanted—but then I thought what good is a compliment from a ten-cent lawyer, and I waived back my recollection. A lawsuit is the suicide of time."

Edison now had other problems. His manufacturing companies employed between 2,000 and 3,000 workers. They owned shares worth some $4 million in the companies they supplied. But they were so short of cash that they were barely able to meet new orders. Edison agreed to a merger with Edison Electric.

The new company was to be called the Edison General Electric Company. It was dominated by the banker J. Pierpont Morgan. Edison Electric exchanged its own shares for shares in Edison General Electric worth $3,500,000. The manufacturing companies picked up slightly less in shares and cash. Of this, Edison himself collected rather more than half. Further capital expansion left him with a one tenth share of Edison General Electric. But he was still a director.

In 1892, competition between Edison General Electric and the rival electrical firm of Thomson-Houston led to a price-cutting war. The cost of bulbs fell from a dollar to forty-four cents. Anxious to avoid further losses, Morgan and Charles A. Coffin, head of Thomson-Houston, decided to join forces in a new firm to be called the General Electric Company.

Edison was outraged. He had invented electric lighting. The fortunes of G.E.C. rested largely on his own patents. Yet Morgan had not consulted him about the new merger. Even his name had been dropped from the title. Shortly afterwards, he sold his shares in G.E.C. and resigned his directorship. He had finished with electric lighting for good.

Meanwhile, even greater changes had taken place in his private life. In 1884, his wife, Mary, had died of typhoid while she was still under thirty. Except at meals and on Sundays she had seen little of her husband because of the long hours he spent in his laboratory. Yet, there was a deep affection between them. Though quiet, she had coped well with her swift rise from factory girl to millionaire's wife.

At thirty-eight, then, Edison found himself a widower with three children. He sent Marion, now thirteen, to a boarding school and arranged for her brothers, Thomas Junior and William Leslie, to live with their mother's sister, Alice, who had married a foreman at Menlo Park.

Edison's friends invited him out to dinner. They took him to concerts and theatres. Early in 1885, while staying at the Woodside, Boston, home of Ezra Gilliland, an old friend who was now working for him, he met Mina Miller, a striking, self-assured girl of eighteen. She was the daughter of a self-made manufacturer of Akron, Ohio. He soon fell in love with her.

"Saw a lady who looked like Mina," he wrote in his

> Cor Ave B & 17th St
> New York Sept 30 1885
>
> My Dear Sir
>
> Some months since, as you are aware, I was introduced to your daughter Miss Mina. The friendship which ensued became admiration as I began to appreciate her gentleness and grace of manner, and her beauty and strength of mind
>
> That admiration has on my part ripend into love, and I have asked her to become my wife. She has referred me to you, and our engagement needs but for its confirmation your consent.
>
> I trust you will not accuse me of egotism when I say that my life and history and standing are so well known as to call for no statement concerning

yself. My reputation is so far made that
recognize. I must be judged by it for
ood or ill.

need only add in conclusion that the step
have taken in asking your daughter
 intrust her happiness into my keeping
as been the result of mature deliberation,
nd with the full appreciation of the
esponsibility I have assumed, and the
uty I have undertaken to fulfil

I do not deny that your answer will
riously affect my happiness, and I
ust my suit may meet with your
pproval.

　　　Very sincerely yours

　　　　Thomas A Edison

o Lewis Miller Esq
　　Akron
　　　Ohio

Above Edison's letter to Lewis Miller, asking for his daughter's hand in marriage. *Above right* Mina Miller Edison—the second Mrs. Edison.

diary. "Got thinking about Mina and came near being run over by a street car. If Mina interferes much more, will have to take out an accident policy."

One night, his friends pointed out the beauty of the full moon on the water. "Couldn't appreciate it, was so busy taking a mental triangleation [*sic*] of the moon, the two sides of the said triangle meeting the base line of the earth at Woodside and Akron, Ohio."

Edison proposed and was accepted. He bought a huge rambling house called Glenmont at West Orange, New Jersey near New York, and had a winter residence built at Fort Myers, Florida. Here he took his bride after a wedding for which the whole of Akron turned out. He had chosen well. Mina bore his long absences patiently. She shone at his side on public occasions. She gave him a daughter, Madeleine, and two sons, Charles and Theodore.

Edison driving the 1882 version of his electric locomotive originally built in 1880. The motor was one of his "Long-waisted Mary Ann" dynamos.

Edison's inventive genius was now blazing less brightly. Compared with his rivals, however, he was still a power to be reckoned with. Ideas never stopped coming. While still working on the elusive light filament, he had found time to build an electric locomotive that worked well over two and a half miles of track at Menlo Park. He did not go on with it because he was too busy with other things. So he missed out on one of America's fastest growing industries.

It was, perhaps, the first sign that he was losing touch with the mainstream of technological progress.

His failure to pursue wireless was another. It was a natural development of telegraphy. Moreover he himself made three of the basic inventions on which Guglielmo Marconi, the Italian pioneer of radio, was to build. These were the microphone, an aerial mast which he used for sending signals through space by means of magnetic induction, and an electronic lamp from which the radio valve was eventually developed.

In the 1890s, he foresaw the possibilities of the motor car. Steam engines were too cumbersome, petrol engines unreliable. The future seemed to lie with electric cars, which were already appearing on the streets in fair numbers. What they needed was a light, durable, high capacity battery to give them an adequate range.

After ten years, he perfected one. It was to prove useful in industry, in radio telegraphy and in ships, especially submarines. It even made a small profit. But Edison missed the huge motor car market. Ironically, it was captured by one of his own engineers who left to develop a petrol engine and later became one of Edison's closest friends. His name was Henry Ford.

Edison also turned his back on new methods of distributing electricity. His own equipment was

FORD CAR PRESENTED TO
THOMAS A EDISON BY HENRY FORD
— 1914 —

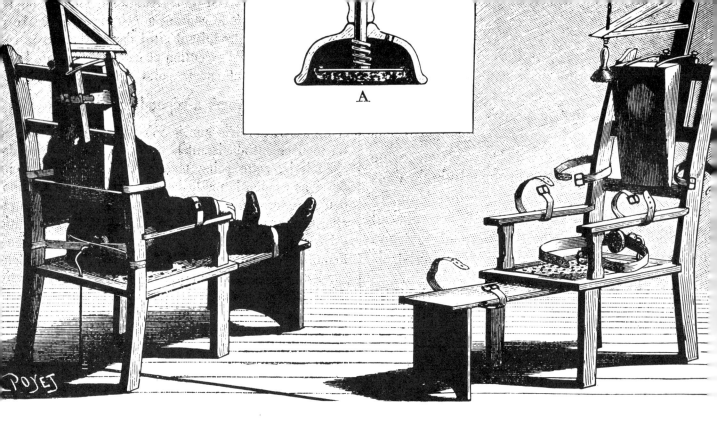

designed for generating and distributing direct current at a voltage of 110. It was safe and it sufficed for his electric lamps. But it was expensive to transmit over long distances and was not powerful enough to work the heavy industrial machinery that was now being developed.

George Westinghouse, founder of the Westinghouse Electrical and Manufacturing Company, plumped for alternating current (a.c.). Even at 5,000 or 10,000 volts, it could be transmitted cheaply over huge distances. It was adequate for industry and could be "stepped down" for ordinary lighting purposes. The system made possible the harnessing of the Niagara Falls as a source of power.

Edison allowed his publicists to sensationalize the alleged dangers of a.c. Dogs were electrocuted at public lectures. Partly as a result, New York State brought in the "electric chair" as a method of executing murderers. A.c. was here to stay.

From 1st January, 1889, New York State used the electric chair to execute criminals in preference to the "barbaric punishment of death by hanging." Edison's arguments against alternating current, which highlighted its dangers, almost certainly helped the introduction of the electric chair, though Edison himself was against the whole idea of capital punishment.

Edison in 1923 sitting in the first electric truck that he had made forty years previously.

Edison's most spectacular failure was in iron ore crushing. The Appalachian mines which had produced high grade ore were worked out. Transport costs made the use of Michigan ore uneconomic for the Eastern ironworks. He developed a method of crushing low grade ore and separating the iron-rich particles from the sand by magnetism. These were then bound into briquettes for ease of handling. He bought or leased land in New Jersey containing an estimated 200 million tons of low grade ore. He built elaborate works at Ogdensburg and hired 500 men. After ten years, he was ready to fill an order for 10,000 tons for the Bethlehem Steel Company.

At this point, a scheme was devised for carrying high grade ore cheaply from Minnesota by rail and boat. Prices dropped sharply. It cost Edison twice as much to produce his briquettes as he could now get from selling them. In 1899, he closed the scheme down. He had lost $2 million and incurred debts of a further $300,000. "We had a hell of a good time spending it," he said.

The iron mine and ore-processing plant at Ogdensburg, New Jersey, in the 1890s—the scene of Edison's most spectacular failure.

7 *Indian Summer: the Phonograph and Moving Pictures*

Edison's main interest was now in research. Having outgrown Menlo Park, he had built himself a large new laboratory half a mile from his home at West Orange, New Jersey. It was a three storey brick building, 250 feet long and 60 feet wide. There were also four single storey buildings each some 100 feet by 25 feet.

Many of the departments changed their names as Edison's interests changed. In the early 1890s, they included a heavy machine shop, a precision department, a mercury vacuum pump room, a lamp test room, a chemical room, a galvanometer department, an ore-milling department and a store room, as well as a lecture hall, an exhibition hall and a library of 40,000 volumes. He was still the go-getting candy butcher at heart. Framed by his desk was the note signed by the editor of the Detroit *Free Press* when he asked for extra copies on the day of the battle of Shiloh: "Give this boy all the papers he wants on credit."

Edison now employed many university-trained scientists. Each project was handled by a separate team. He supervised them all personally. Improvements to his lamps and lighting system were regularly made. Two other schemes eventually made him a multimillionaire.

For ten years he regarded the phonograph as a scientific novelty with few commercial possibilities. In 1886, however, Alexander Bell, the inventor of the telephone, patented an improved machine with a wax-coated cylinder. He called it a "graphophone". Stung into competition, Edison switched over to a solid wax cylinder. He devised a floating stylus and an electroplated "master" record from which a large

Alexander Graham Bell's
"graphophone"—a much im-
proved version of Edison's
phonograph using a wax-coated
cylinder.

number of copies could be pressed. A constant speed motor was added. By 1888, he had a saleable machine.

As so often happened, he proved to be a better inventor than businessman. Jesse W. Lippincott, a Pittsburgh glass manufacturer, had already bought up the Bell patents. He now bought up Edison's too. Edison himself received half a million dollars and the sole right of making the phonographs.

Later, he found that Lippincott had paid Gilliland, now Edison's business manager, a further $250,000 for the right to sell the machines. Gilliland paid a third of this to Edison's lawyer, John Tomlinson. The selling rights were worth, perhaps, $50,000. The other $200,000 was a concealed bribe in return for which Gilliland and Tomlinson had persuaded Edison to sell the patent rights for half their true value.

By the time Edison found out, the conspirators had decamped for Europe. Lippincott failed to make a

Edison's improved phonograph of 1895 which had a spring motor.

success of the business and was soon paralysed by a stroke. Edison was owed so much for machines that he was able to take over the company, thus regaining control of his patents.

He developed the business slowly but successfully. At first, he thought his two-minute *cylinders* would be mainly used for office dictation. But the Victor and Columbia companies showed that there was now a huge market for two- and four-minute entertainment *discs* made for the Victor talking machine or gramophone invented by Emile Berliner. So Edison too recorded first popular, then serious music and later, he too switched to discs.

Meanwhile he had also been attracted by the idea of motion pictures. "In the year 1887," he said, "the idea occurred to me that it was possible to devise an instrument which should do for the eye what the phonograph does for the ear."

It was well known that the image of an object seen

The first "Kodak" camera of 1888, with a packet of the first commercial celluloid rollfilm of 1889.

by the eye persisted in the brain for up to a tenth of a second after the object itself was no longer in sight. So a series of still photographs, showing successive positions of a body in motion, would give an appearance of movement if they could be shown in rapid succession.

Ordinary cameras then used glass plates which were far too big and clumsy. After a number of failures, Edison found that tiny pictures could be photographed spirally on a celluloid film wrapped round a phonograph-type cylinder. As it turned, sparks from an induction coil lit them up one by one. Viewed through a magnifying glass, they then gave the illusion of motion. An early effort showed an assistant in a white sheet, waving his arms about.

George Eastman (left), inventor of the "Kodak" camera, seen here with Edison in 1928. Edison is standing behind one of his movie cameras.

The breakthrough came when George Eastman, the inventor of the "Kodak," made continuous strips of celluloid film that could be run through a camera. "Allowing 46 exposures per second, as we did at first," said Edison, "we had to face the fact that the film had to be stopped and started again after each exposure. Now, allowing 1/100 part of a second for every impression that was registered, you can see that almost half of our time was gone, and in the remainder of the time we had to move the film forward the necessary distance for the next exposures. . . . All this had to be done with the exactness of a watch movement."

Edison's movie camera, which he called a kinetograph, was a big success. So was the

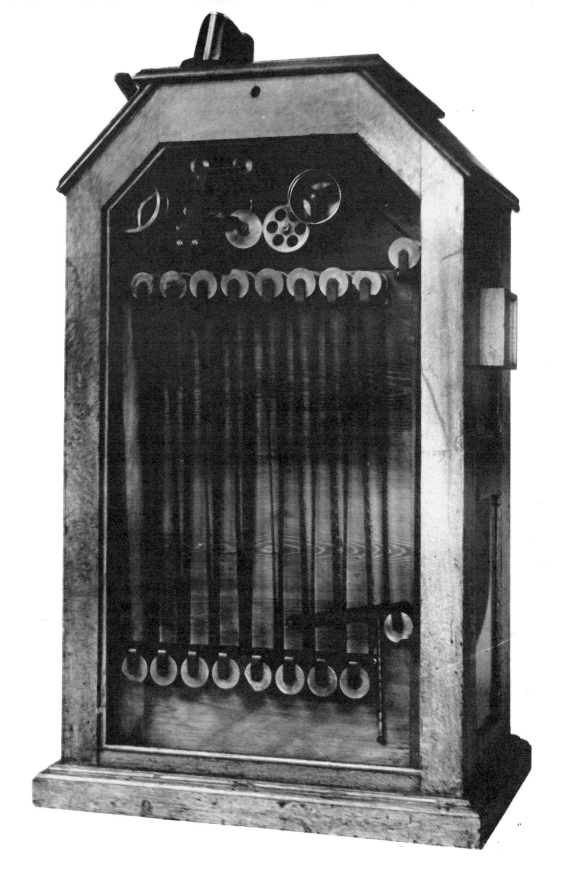

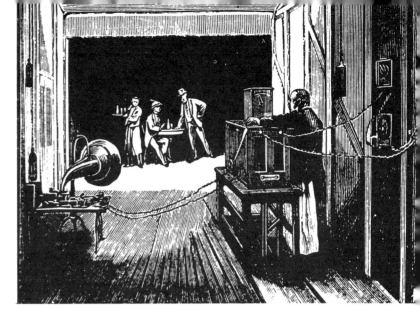

Left Edison's kinetoscope. The moving strips of film were viewed through the eyepiece at the top.

Above The kinetoscope and phonograph combined and *above right* Edison's first experiments with this combination in the late 1880s.

"kinetoscope," a coin-in-the-slot peep-hole machine for showing his films. It had a magnifying glass and a revolving shutter with a hole in it. This revealed the pictures one at a time in rapid succession. Kinetoscope parlours blossomed all over America.

Edison made films for them in the "Black Maria," a fifty-foot, black-lined studio with a retractable roof. It could be swung round on a pivot to catch the sun at any time of the day. Boxers, dancers, jugglers and performing bears appeared in films which lasted rather longer than a minute.

If Edison had put all his energies into motion pictures, he might well have won a world monopoly. But he was busy with too many other things. He did not patent the kinetograph and kinetoscope until 1891, two years after inventing them. Even then, he did not take out European patents because, he said, they weren't worth the $150 fee! He was also slow to develop a system of projecting films on to a screen.

The result was inevitable. He made steadily more elaborate films, culminating in thousands of fourteen-minute dramas and comedies. He built a $100,000 studio in the Bronx district of New York. His manufacturing companies reaped huge profits from making equipment.

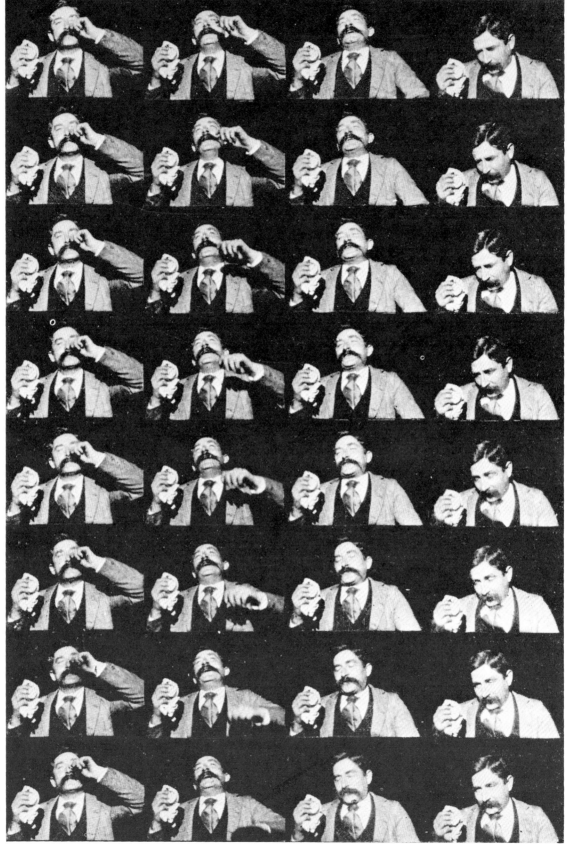

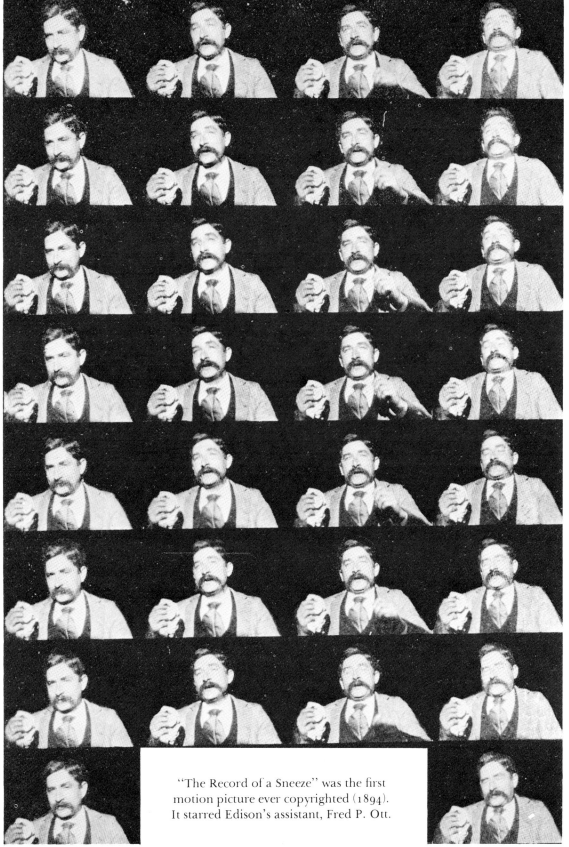

"The Record of a Sneeze" was the first
motion picture ever copyrighted (1894).
It starred Edison's assistant, Fred P. Ott.

Yet it was only a partial success. His delays had allowed rivals to catch up with him. European manufacturers freely copied his unpatented inventions, developed them and sold the rights in America. Edison once more became involved in a tangle of patent suits which took sixteen years to resolve. To save further trouble, all the parties involved then pooled their patents. Edison's share of the royalties amounted to $1 million a year.

The pool infringed American laws against such arrangements. In 1917, it was broken up. By then, Edison was a national legend, wealthy, and loaded with honours. Though devoted to his second wife, Mina, he had no interests outside his work. Henry Ford was one of his few intimate friends.

Above Among the many scientists and inventors who visited Edison in the U.S.A. was Rudolf Diesel, inventor of the diesel engine, seen here in the study of Edison's home at West Orange, New Jersey in 1912.

Edison showing his phonograph to Governor A. Harry Moore of New Jersey in the late 1920s.

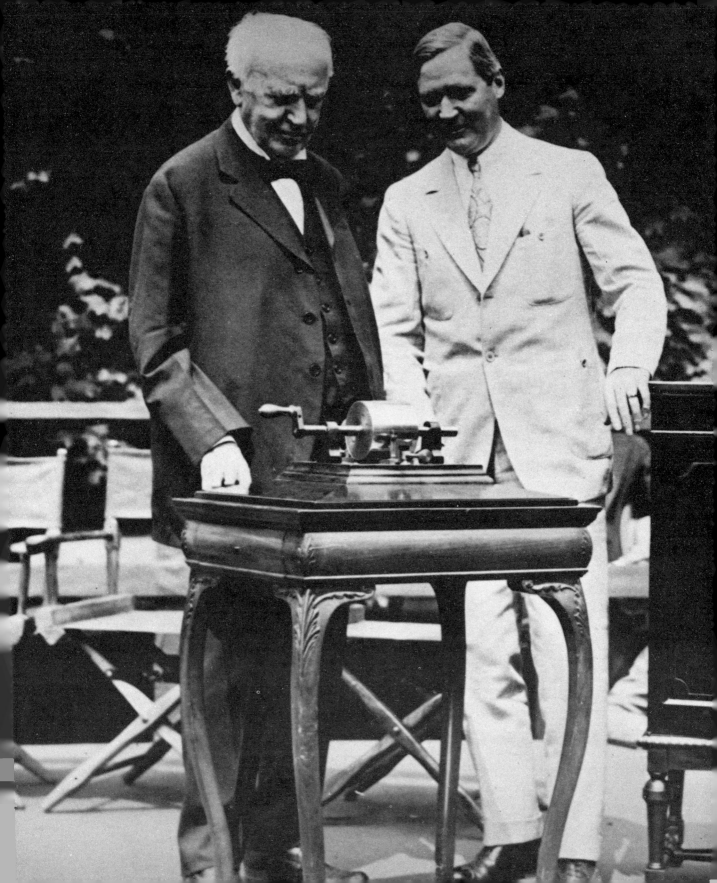

During the First World War, he became head of a board of technologists formed to advise on military inventions. He himself thought up several devices for fighting the submarine menace. He set up factories for making benzol, carbolic acid and other chemicals in short supply, and made yet another fortune.

In his later years, he was stocky, but still healthy and active. His hair was white and untidy, his face impassive under bushy eyebrows. But his judgment steadily worsened. He refused to have anything to do with radio receivers for the popular market. He thought they had no future. He changed his mind in 1928 but the slump cut demand. In 1931, production was stopped after the Edison companies had lost $3 million. A year before, they had stopped making phonographs and records. Edison had stubbornly refused to keep up with improved methods of sound reproduction.

He spent years trying to find a cheap rubber substitute which could be grown in America. By 1929, after investigating some 15,000 plants, he had devised a scheme for extracting liquid rubber from goldenrod. But it was dearer than imported rubber. Moreover, synthetic rubber had already been invented in Germany.

Edison was almost eighty when he resigned from his companies. His sons by his first wife both proved a disappointment. Thomas junior became involved in shady business ventures and committed suicide in 1936. William Leslie, after serving in the American army, eventually became a gentleman farmer and died in 1941.

His sons by Mina, however, were all he could have wished. Charles became the highly successful head of the Edison business interests. Theodore became technical director of the Edison Laboratory but eventually left to set up as a scientific consultant on his own.

In 1929, Henry Ford arranged a celebration at

Expo 1915 in San Francisco featured many of Edison's inventions. Here he is discussing his telegraph perforator with Henry Ford.

Dearborn, Michigan for the fiftieth anniversary of Edison's inventing the electric lamp. Ford had shipped in all the surviving equipment and had built a replica of Menlo Park laboratory. President Herbert Hoover and Madame Marie Curie headed a swarm of notables who had come from all over the world to pay homage to Edison. As Hoover was finishing his speech, Edison collapsed.

Doctors brought him round. But a number of illnesses had taken hold of his ageing frame. He had gastric ulcers, kidney disease and diabetes. He gradually gave up his laboratory work, spending his days sleeping or reading in the garden of his West Orange home. He would allow only Mina to take care of him. He insisted on measuring out his own medicine and on studying his own blood tests through a microscope on his bed. On 17th October, 1931, he fell into a coma. The next day he died, at the age of eighty-four.

Edison was a giant. He had gigantic successes and gigantic failures. He had a giant's zest, a giant's power of recuperation and, until his last years, a giant's vision. If he had invented only the electric light bulb, he would have been noteworthy. That he also gave us the microphone, the kinetoscope, the phonograph and scores of other devices makes him one of the greatest inventors the world has ever known.

Most important, he pioneered the industrial research laboratory, thus making possible the technological progress which enabled us to put a man on the moon. Whether he would have thought such a project worthwhile is doubtful. But once it was achieved, he would have been ready with a dozen ideas for turning it to good advantage. Making invention profitable was his main aim in life.

Thomas Edison on his eighty-first birthday "clocking in" at his laboratory.

Date Chart

1847	Thomas Alva Edison born in Milan, Ohio on 11th February.
1854	Edison family moves to Port Huron, Michigan.
1859	Becomes "candy butcher" on the Detroit train.
1860	Abraham Lincoln elected President of the U.S.A.
1861–5	American Civil War.
1863–9	Edison works as a telegraphist.
1868	Applies for his first patent on a vote-recording machine.
1869	Resigns from Western Union, Boston to become full-time inventor.
	Applies for patent on stock ticker.
	Moves to New York.
	Joins Gold Indicator Company.
	Leaves to set up as an "electrical engineer".
	Marries his first wife, Mary.
1870	Western Union buys up his stock ticker.
1871	Sets up as a manufacturer of stock tickers.
1872	Applies for patent on automatic telegraph.
1873	First of several visits to England.
1874	The multiplex telegraph perfected.
	The motograph circumvents Page relay patents.
1875	Swindled by financier Jay Gould.
1876	Builds Menlo Park research laboratory.
1877	Perfects telephone transmitter (microphone).
	Applies for patent on phonograph.

1878	Applies for patent on first electric lamp. Edison Electric Light Company set up.
1879	Applies for patent on vacuum lamp with platinum filament. Applies for patent on carbon filament lamp.
1880	Applies for patent on new method of extracting ore. Lighting plant installed in *S.S. Columbia*. First run of electric train at Menlo Park. Edison Electric Illuminating Company of New York set up.
1880–1	Sets up his own manufacturing companies.
1881	Moves his headquarters to New York. Pearl Street site bought for New York central station.
1882	Lights switched on in New York.
1884	Death of Mary, his first wife.
1886	Marries his second wife, Mina.
1887	Builds new laboratory at West Orange, New Jersey.
1888	Improved phonograph perfected. His manufacturing companies merge with Edison Electric to form the Edison General Electric Company.
1889	The kinetograph and kinetoscope invented.
1890–99	Ore-crushing fiasco in Ogdensburg, New Jersey.
1891	Edison General Electric merged with Thomson–Houston to form the General Electric Company. Edison ousted. Applies for patents on kinetograph and kinetoscope.
1908	Motion Picture Patents Corporation formed as patent pool.
1909	Nickel-iron-alkaline storage battery perfected.

1911	Edison's own companies combined in Thomas A. Edison Inc.
1914–18	First World War.
1914	Fire destroys seven of his factories at West Orange.
1915	President of the Navy Consulting Board (for inventions).
1916	Starts work on anti-submarine devices.
1917	America enters the war.
1922	Nominated "greatest living American" in *New York Times* poll.
1927–9	Develops a natural substitute for rubber.
1928	Awarded Congressional Medal of Honour.
1929	Severe illness—diabetes and kidney disease.
1931	Dies on 18th October.

Further Reading

Easily the best full length biography is *Edison* by Matthew Josephson (Eyre. London, 1961).

Others are:

The Life and Inventions of Thomas Alva Edison by W. K. L. Dickson and Antonia Dickson (Chatto. London, 1894).

Thomas Alva Edison by Francis Arthur Jones (Hodder. London, 1907).

Edison by William Adam Simonds (Allen and Unwin. London, 1935).

Additional material can be found in:

My Friend Mr. Edison by Henry Ford with Samuel Crowther (Benn. London, 1930).

The Diary and Sundry Observations of Thomas Alva Edison edited by Dagobert D. Runes (Philosophical Library. New York, 1948).

Glossary

Arc light   Light produced when electric current passes from one terminal to another.

Circuit   The path of an electric current.

Crank   A handle for turning a shaft.

Diaphragm   A vibrating disc, as in a telephone or microphone.

Duplex   A telegraph circuit that can transmit two messages at once.

Dynamo   A machine for making electricity.

Electromagnet   Iron or steel magnetized by an electric current passing through a coil of wire surrounding it.

Filibuster   Unnecessarily long speech delivered to hold up business in an assembly.

Fuse   Piece of wire that melts when overloaded. It is placed in a circuit to stop dangerously high current causing damage.

Incandescent light   Light produced when electric current causes a filament to glow.

Induction coil   Device for increasing the voltage of an electric current.

Mains   Principal wires in distribution of electricity.

Microphone   Device for changing sound waves into electrical energy.

Morse code   Dot-and-dash system of representing the letters of the alphabet. Devised for telegraphists by U.S. inventor Samuel Morse (1791–1872).

Parallel   A ladder-like arrangement of electric circuits in which all the "rungs" have their positives connected to one upright and their negatives to the other.

Patent   The exclusive right to make, use or sell an invention for a given number of years.

Pirate   To infringe a patent without permission.

Radio   Wireless telegraphy or telephony.

Resistance   Opposition to the passage of an electric current.

Royalty   Payment for use of someone else's patent.

Series   Necklace-like arrangement of circuits in which the positive of one electrical instrument (e.g. a light bulb) is joined to the negative of the next.

Telegraph   Device for sending messages to a distant point, especially by making and breaking an electric circuit in a connecting wire.

Telephony   Method of transmitting sound to a distant point by electrical means.

Transformer   Device for changing the voltage of an electric current.

Volt   Measure of electromotive force, i.e. electrical "pressure."

Voltmeter   An instrument for measuring voltage.

Index

Picture Credits

The author and publishers wish to acknowledge their thanks to the following for permission for illustrations to appear on the following pages:
Kodak Limited, 76, 77 ; The Mansell Collection, *frontispiece,* 11, 21, 52–53 ; Mary Evans Picture Library, 10, 14–15, 25, 34–35, 37 ; Maschinenfabrik Augsburg Nürnberg, 82 ; Paul Popper Limited, 69 ; Post Office, 35 ; Radio Times Hulton Picture Library, *jacket,* 37 58, 71, 84, 87 ; Science Museum, 44, 51, 78, 79 ; Teleprocessing Industries Inc., 17, 26, 30, 31, 85. The remaining pictures are the property of the Wayland Picture Library.